T·H·E FAST A·N·D SLOW ANIMAL B·O·O·K

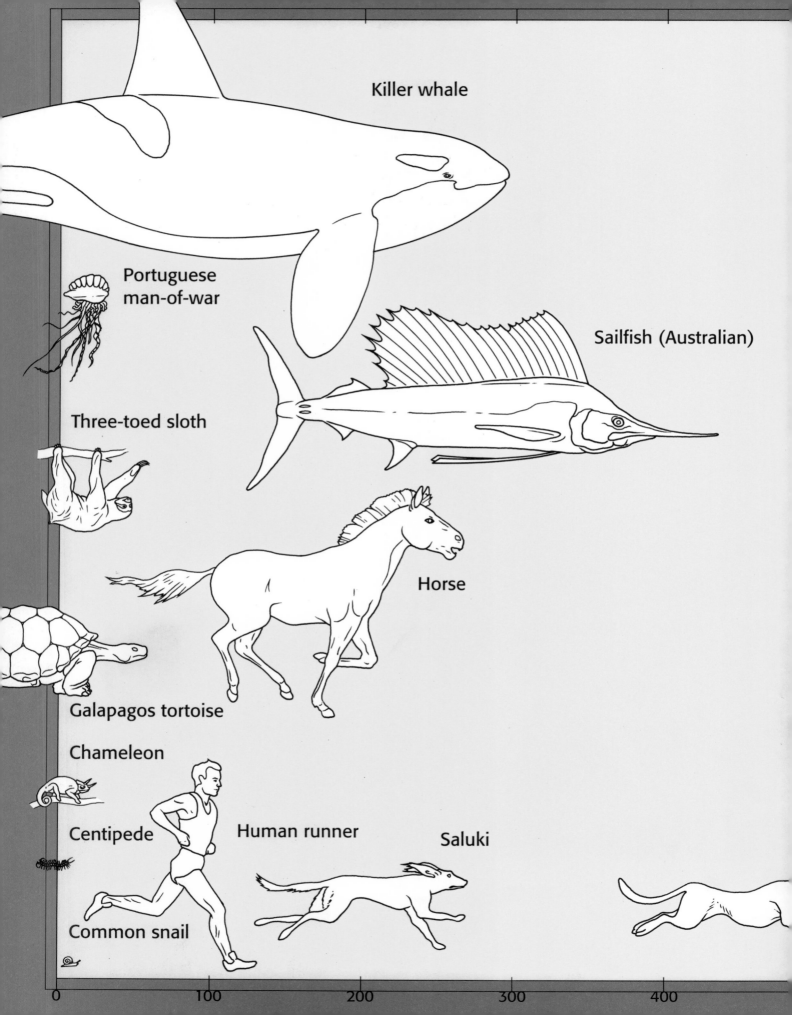

Killer whale

Portuguese
man-of-war

Sailfish (Australian)

Three-toed sloth

Horse

Galapagos tortoise

Chameleon

Centipede

Human runner

Saluki

Common snail

0   100   200   300   400

SPEED BOX

A 40 millimeter box equals 100 meters.
When the diving Peregrine falcon has
flown 3,281 feet (1,000 meters), the
other animals have traveled this far.

Peregrine falcon

Spine-tailed swift

Cheetah

0	600	700	800	900	1000m

Animal
Opposites

T·H·E
FAST
A·N·D
SLOW
ANIMAL
B·O·O·K

DAVID TAYLOR
ILLUSTRATED BY
PETER MASSEY

RSVP
RAINTREE
STECK-VAUGHN
P U B L I S H E R S
The Steck-Vaughn Company

Austin, Texas

Contents

Words found in **bold** are explained in the glossary on page 31.

Published by Raintree Steck-Vaughn Publishers, an imprint of Steck-Vaughn Company.

Illustration copyright © Peter Massey 1994
Text copyright © David Taylor 1994

Editor: Jill A. Laidlaw
Science Editor: Tracey Cohen
Designer: Julie Klaus, Frances McKay
Electronic Production: Scott Melcer
Consultant: Steve Pollock

**Library of Congress
Cataloging-in-Publication Data**

Taylor, David, 1934–
    The fast and slow animal book / David Taylor; illustrated by Peter Massey.
        p.   cm. — (Animal opposites)
    Includes index.
    ISBN 0-8172-3951-0
    1. Animal locomotion — Juvenile literature.
[1. Animal locomotion.] I. Massey, Peter, ill.
II. Title. III. Series: Taylor, David, 1934– Animal opposites.
QP301.T326   1996
591.1'852—dc20   95-8331
                  CIP   AC

Printed in Hong Kong
Bound in the United States
1 2 3 4 5 6 7 8 9 0 LB 99 98 97 96 95

Fast and Slow Animals

We are part of a universe in which everything, from the smallest to the biggest, is moving. The time it takes for something to move a certain distance is called speed. Usually, speed is measured in miles or kilometers (km) per hour.

Are Humans Fast or Slow?

Speed is important in many areas of our lives. The Concorde jet cruises at 1,327 miles (2,137km) per hour. To go into **orbit** up above the Earth, a rocket must reach a speed of 17,885 miles (28,800km) per hour. People design and build these machines. But we creep along compared to them when we move under just the power of our own bodies. We walk at about 4 miles (6km) per hour. Even a champion runner can reach only 30 miles (48km) per hour during a race.

The Animal Kingdom

In the animal kingdom creatures of all kinds move at different speeds. Some don't move at all, and others are very fast. Some, like oysters, move at certain times in their lives and not at others. Why are there these differences in the speeds of animals? A great deal depends upon their lifestyles. If you are a hunter who eats fast-moving animals, you need to be even faster to catch them. If you are one of the hunted animals, then moving quickly could save your life.

On the other hand, not moving can make a lot of sense. The eyes of some hunters, like snakes, do not see animals that stay still. Sea anemones wander slowly through the water because their dinners are delivered to their waiting tentacles by that same water. So, there is no need to go hunting!

What Is Best?

There are advantages and disadvantages to being slow or fast. Fast animals burn up lots of energy and need more food. Very slow animals like turtles do not use up much energy. But they do need the defense of their hard shell to avoid being gobbled up. So, very fast, very slow, or in-between animals do what they do at the speeds that are best for them.

Where in the World?

You can see the places where the animals talked about here live. Turn to pages 30 and 31 in the back of this book.

How Do We Measure Distance?

When we measure length, we use either standard or metric measure. In standard measure we use inches, feet, yards, and miles to see how far something travels. In metric measurement, we talk about centimeters, meters, and kilometers. If you look at the ruler on the bottom of these pages, you will see both systems of measurement.

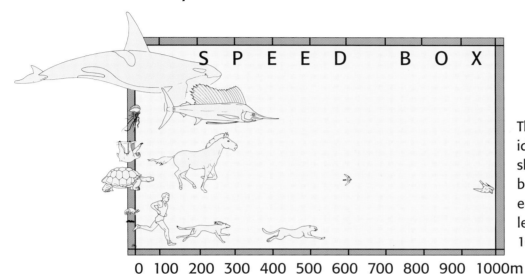

These speed boxes give an idea of how fast and how slow the animals in this book are in comparison to each other. A 1 centimeter length represents 100 meters.

0 100 200 300 400 500 600 700 800 900 1000m

The man in the bottom left corner of the box is running at a very good speed for a distance of 100 meters. He is running almost at the speed of a record-breaking champion.

How Do We Measure Speed?

When measuring speed we record how long it takes for someone or something to travel a certain distance. The distance can be anything from inches and feet to miles or from centimeters and meters to kilometers. We measure the time in seconds, minutes, hours, days, months, and years.

9 10 11 12 13 14 15 16

22 23 24 25 26 27 28 29 30 31 32 33 34 35 36 37 38 39 40 41 42

The Sloth

Imagine an animal that moves so slowly algae grow in its hairy coat! It spends most of its time hanging upside down in the trees. When it moves, it climbs along branches, holding on with its long, curved claws. It moves at around 1 mile per hour (mph) (2kmph). On the ground it crawls even more slowly, dragging itself on its belly at about 0.06 to 0.09 mph (0.09 to 0.14kmph).

   Its name in English means "idle" and comes from the old word for "slow." It is the sloth.

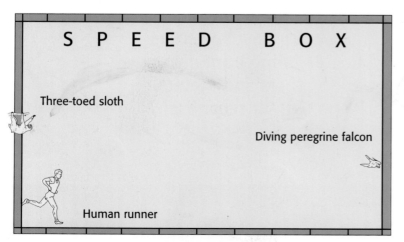

S P E E D   B O X

Three-toed sloth

Diving peregrine falcon

Human runner

0 100 200 300 400 500 600 700 800 900 1000m

Three-toed sloth

Saving Energy

Sloths have a lower body temperature than most **mammals**. This, together with their unhurried lifestyle, helps them to save energy.

6

A Nasty Bite

Sloths may seem defenseless, but they can bite and inflict serious wounds with their claws. Their biggest enemies are people because people destroy the rain forest. This loss of the sloths' home means that one type of sloth, the maned sloth of southeast Brazil has become rare.

Eating in the Trees

Sloths eat leaves that few other animals want and quietly go about their business in the forests. Their gray-brown fur is tinted green by algae. Algae are primitive plantlike, living things that grow on the fur and give it camouflage.

After all this, you must think the sloth is a failure. But it is not so. Sloths, of which 5 **species** exist, are unique and highly successful creatures. In some parts of Central and South America, sloths make up to a quarter of the total weight of mammals in the countryside.

The Chameleon

The chameleon is a lizard, which is a reptile. This means that they breathe air and have backbones and scaly skin. The chameleon moves almost as slowly as the turtle. See pages 24-25. Chameleons can move very quickly if they want to, but they stay still for long periods of time to protect themselves. True chameleons live only in Africa, Madagascar, Arabia, South India, Sri Lanka, Crete, and southern Spain. There are about 85 kinds of chameleons. Creeping along slowly and staying still are what the chameleon does best. The chameleon lies in wait for flies and other insects to capture and eat. A chameleon's skin is usually colored gray-green to blend in with the leaves it lives in. But the chameleon's skin can change color, becoming deep brown or creamy, and patterns of yellow spots or dark bars can appear and disappear.

A Killer Tongue

Although the chameleon is a very slow mover, its tongue is not. Equipped with muscles and an elastic root, its long tongue can shoot out rapidly to capture insects, spiders, and other small animals. On the tip of its tongue is a sticky mucus that traps the **prey**. Quick as a flash the tongue and the prey glued to it, are reeled in like a very fast fishing line.

Amazing Eyes

Chameleon's eyes are covered by eyelids that are joined together so that only the pupil shows through a small central hole. Their eyes move independently and can both focus on one place. These amazing eyes are useful for judging distances so that the chameleon's tongue hits its target.

S P E E D   B O X

Chameleon

0 100 200 300 400 500 600 700 800 900 1000m

Jackson's chameleon

The Centipede

Animals use legs for walking and running. You might think that the more legs you have, the faster you can run. This is not always true, but it is for most centipedes. Depending on the species, adult centipedes have 15 to 177 pairs of legs that sweep them along on waves of movement. The 1-inch (2.54-cm) long house centipede has been clocked at a shocking 16.5 inches (42cm) per second over short distances.

Centipedes belong to a group of animals called **chilopods**. These animals do not have backbones. Their bodies, which are divided into many segments, have a hard outer shell. They breathe through little tubes that carry air to all parts of their bodies.

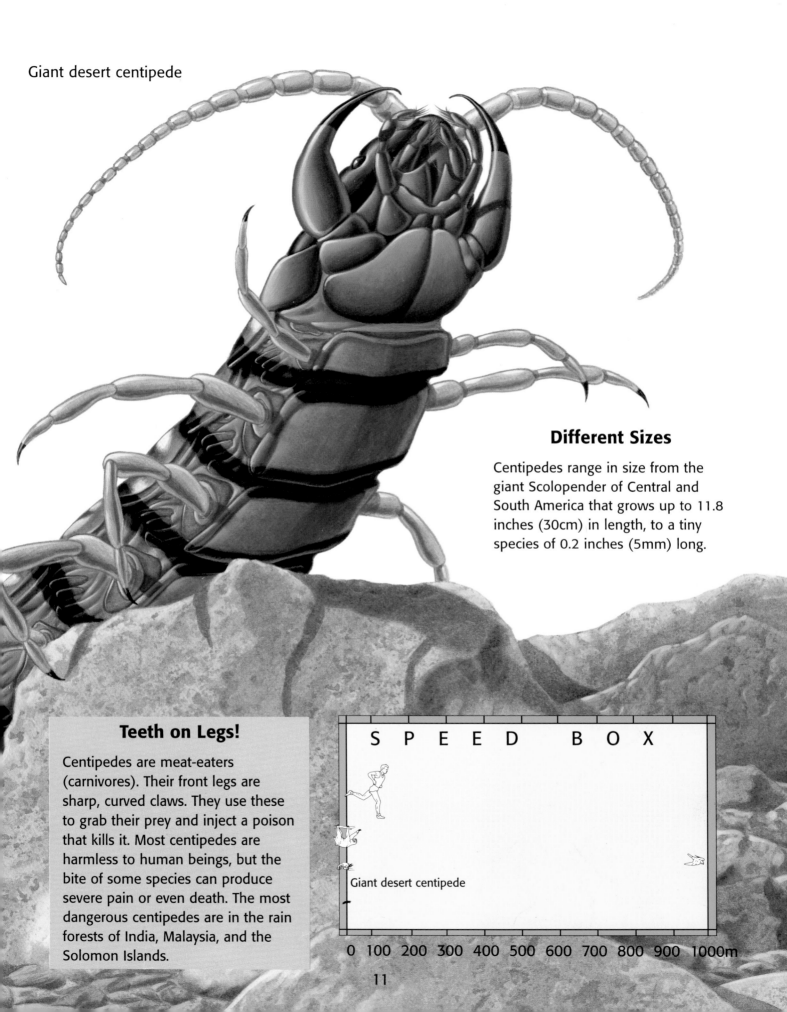

Giant desert centipede

Different Sizes

Centipedes range in size from the giant Scolopender of Central and South America that grows up to 11.8 inches (30cm) in length, to a tiny species of 0.2 inches (5mm) long.

Teeth on Legs!

Centipedes are meat-eaters (carnivores). Their front legs are sharp, curved claws. They use these to grab their prey and inject a poison that kills it. Most centipedes are harmless to human beings, but the bite of some species can produce severe pain or even death. The most dangerous centipedes are in the rain forests of India, Malaysia, and the Solomon Islands.

S P E E D   B O X

Giant desert centipede

0 100 200 300 400 500 600 700 800 900 1000m

The Cheetah

The ace sprinter among wild animals is the cheetah. This handsome spotted cat, with its long legs and tail, is the fastest of all mammals on land. Normally, when hunting, the cheetah dashes along at 72 mph (120kmph). But the big cat can keep up that pace for only a short distance—just 960 to 1,300 feet (300 to 400m). The average chase is 550 feet (170m) and lasts about 20 seconds.

The cheetah is perfectly built for a short-distance, high-speed hunting cat. The cheetah kills quick-moving animals like small antelopes. However, if it cannot catch one in its first whirlwind dash, it will soon give up the chase.

Treated Like Royalty

In the past, cheetahs were called hunting leopards. Kings in Europe and the Middle East tamed these cheetahs imported from Africa for hunting. There are pictures of cheetahs being carried out to the hunting fields on horseback, seated on a pillow behind their master's saddle.

Cheetah

A Power-Packed Body

The cheetah has long legs and a long, flexible back. Strong muscles in its thighs, back, and shoulders help it run. As it gallops, the cheetah arches its back and springs with both back and front legs. It uses its tail for balance. The cheetah's claws, more like a dog's claws, stick out and grip the ground. They are like spikes in track shoes. The grooves on the pads of its feet keep it from skidding.

SPEED BOX

Cheetah

0 100 200 300 400 500 600 700 800 900 1000m

The Horse

Speed and horses have always been linked. The Pony Express carrying mail across the United States in the last century is just one example. Horses also have power. They have been used to carry loads and pull wagons. When cars began to replace horses, "horse power" was still used to describe an engine's power. A 10 horse-power engine can do the work of 10 horses.

Horses are descended from a small doglike animal called *Hyracotherium*. This animal lived about 60 million years ago. Hyracotherium, with three toes on its hind feet and two toes on its front feet, lived in forests and ate leaves. It carried its weight on doglike foot pads. Over 60 million years, Hyracotherium's horselike descendants gradually lost toes until there was just one toe left on each foot. All the power of the horse's muscles is put into these single toes. The nails of the toes have grown bigger and hardened to become hooves ideal for running.

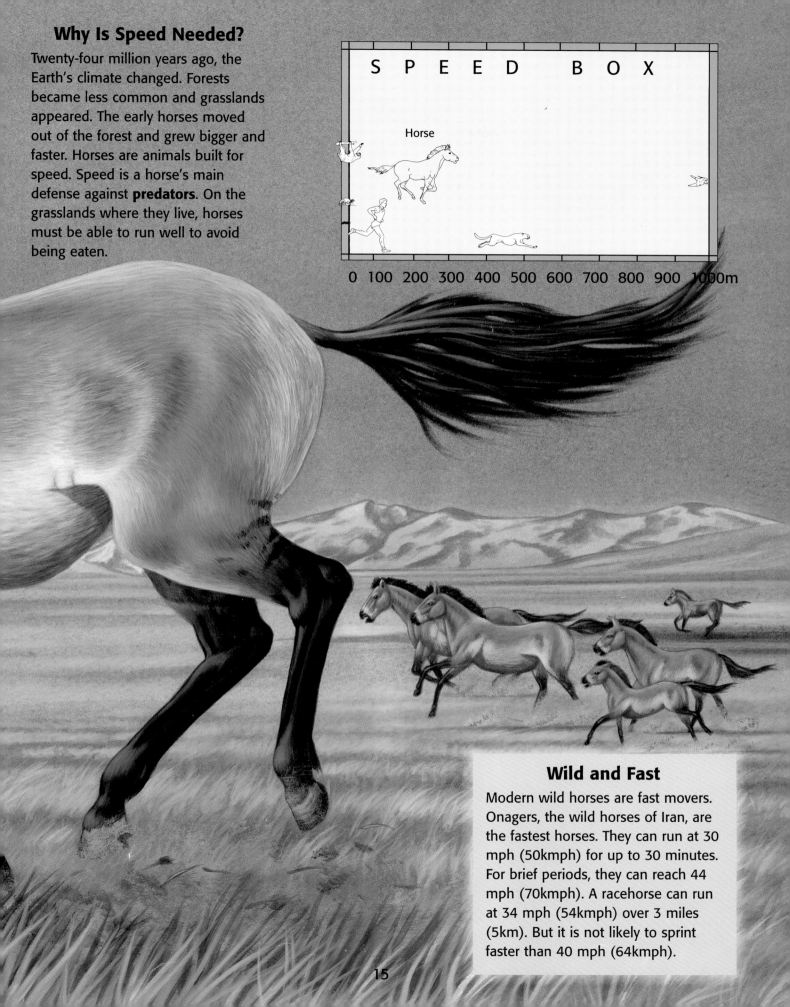

Why Is Speed Needed?

Twenty-four million years ago, the Earth's climate changed. Forests became less common and grasslands appeared. The early horses moved out of the forest and grew bigger and faster. Horses are animals built for speed. Speed is a horse's main defense against **predators**. On the grasslands where they live, horses must be able to run well to avoid being eaten.

S P E E D   B O X

Horse

0 100 200 300 400 500 600 700 800 900 1000m

Wild and Fast

Modern wild horses are fast movers. Onagers, the wild horses of Iran, are the fastest horses. They can run at 30 mph (50kmph) for up to 30 minutes. For brief periods, they can reach 44 mph (70kmph). A racehorse can run at 34 mph (54kmph) over 3 miles (5km). But it is not likely to sprint faster than 40 mph (64kmph).

15

The Portuguese Man-of-War

There are some animals that are unable to move around under their own power. Instead, they make use of the power of wind and water. The Portuguese man-of-war is one such animal. It is a kind of jellyfish with an inflated, or blown-up, blue balloon on top. The balloon is filled with gas. It allows the jellyfish to float at the surface of the water. The balloon may be 4 to 12 inches (10 to 30cm) long. Stinging **tentacles**, long, thin limbs, hang down from the balloon up to 100 feet (30m) deep in the water. The tentacles trap fish and other prey.

Jellyfish are soft-bodied animals without a backbone. The Portuguese man-of-war belongs to a group of jellyfish called siphonphores. Each of these animals is made up of many small individuals. They cling together in a colony or group.

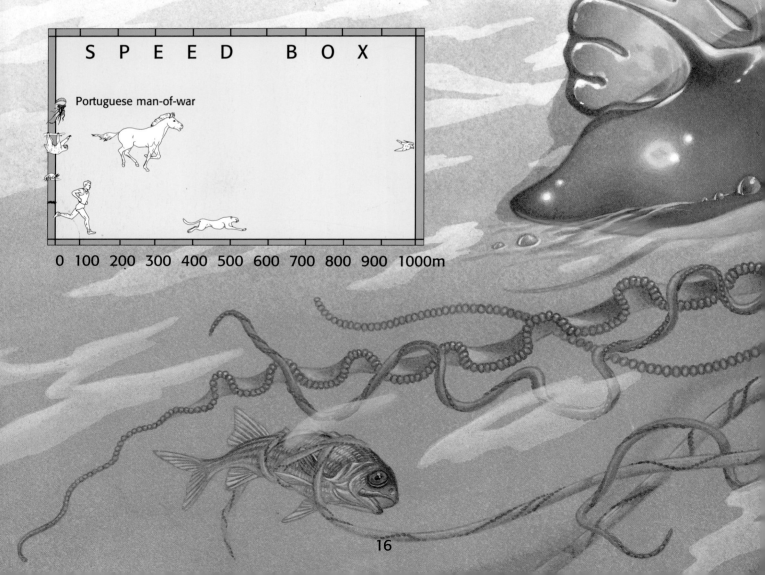

SPEED BOX

Portuguese man-of-war

0 100 200 300 400 500 600 700 800 900 1000m

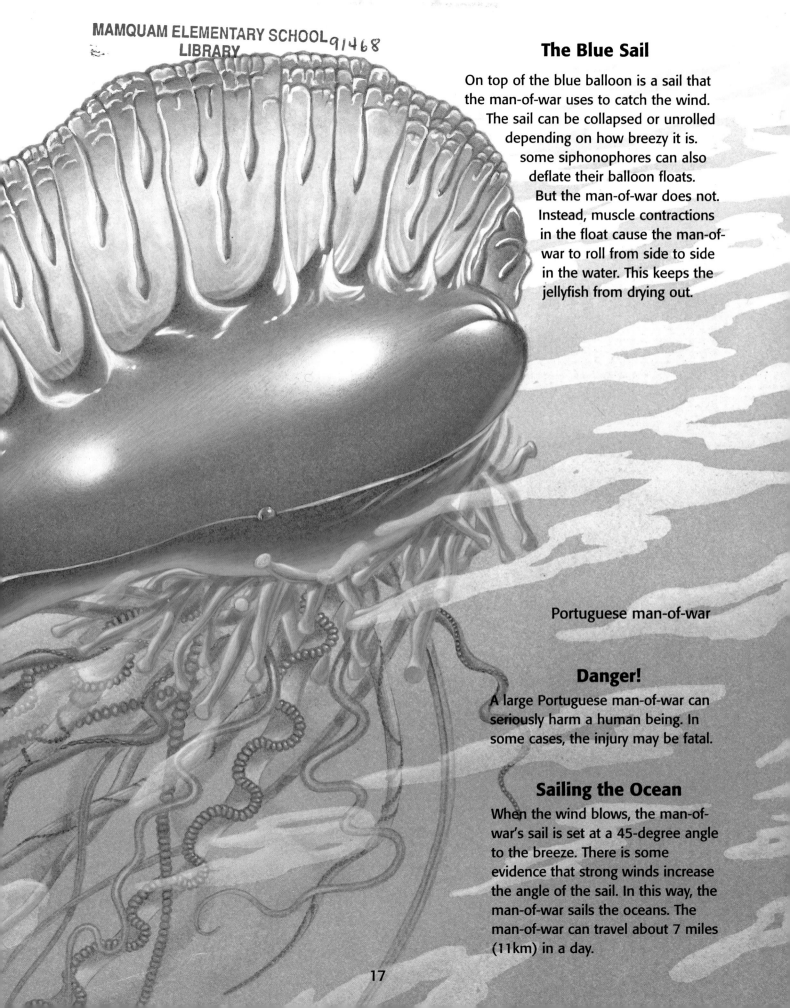

The Blue Sail

On top of the blue balloon is a sail that the man-of-war uses to catch the wind. The sail can be collapsed or unrolled depending on how breezy it is. some siphonophores can also deflate their balloon floats. But the man-of-war does not. Instead, muscle contractions in the float cause the man-of-war to roll from side to side in the water. This keeps the jellyfish from drying out.

Portuguese man-of-war

Danger!

A large Portuguese man-of-war can seriously harm a human being. In some cases, the injury may be fatal.

Sailing the Ocean

When the wind blows, the man-of-war's sail is set at a 45-degree angle to the breeze. There is some evidence that strong winds increase the angle of the sail. In this way, the man-of-war sails the oceans. The man-of-war can travel about 7 miles (11km) in a day.

17

The Saluki

Humans domesticated dogs between 11 and 12 thousand years ago. Dogs were first used for hunting. Then they were trained for herding sheep and goats. As dogs spread throughout the world, their traits changed. In cold climates, dogs develop thick, warm coats. In the desert, they grew tufts of hair that protected their feet against hot sands. Humans also bred dogs for certain traits like speed and strength.

Saluki

The Saluki is one of 400 breeds of **domestic** dogs. It is not as fast as the greyhound (see right) over short distances. But it has more **stamina** and is still used by the **Bedouin** to hunt. Its top speed is probably about 39 mph (62kmph).

The Egyptians hunted with falcons on their wrists and salukis on leads. See pages 26-27. **Mummified** salukis have been found in Egyptian tombs.

Sight Hounds

In the Middle East, sleek, long-legged dogs with long, flexible bodies were chosen to hunt game in open desert country. "Sight hounds" or "gaze hounds" rely on good eyesight during the chase. The descendants of these dogs are breeds such as the greyhound, saluki, and Afghan hound.

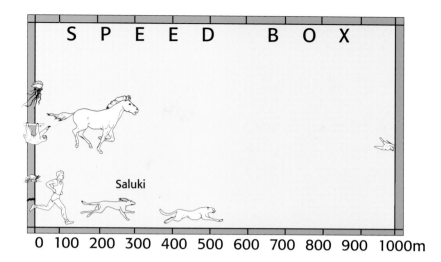

S P E E D   B O X

Saluki

0 100 200 300 400 500 600 700 800 900 1000m

The Fastest

The fastest modern dog is the greyhound, which has been recorded running at a speed of just over 45 mph (72kmph).

19

The Sailfish

It is not easy to measure the speed of fish. Many of them have streamlined bodies. These allow them to move through the water very quickly.

The champion short-distance swimmer among fish is the sailfish. With its large saillike fin, the fish can sprint at over 67 miles (108km) an hour. The sailfish cuts down water resistance by pressing some of its fins tight against its body. It folds the dorsal fin into a slot running along its back. The sailfish is the "cheetah" (see pages 12-13) of the oceans because it uses its speed to hunt other fish.

Pacific sailfish (Australian)

S P E E D   B O X

Sailfish

0 100 200 300 400 500 600 700 800 900 1000m

The Biggest Sailfish

Sailfish are found in all tropical waters. They can be 11 feet (3.4m) or more and weigh over 200 pounds (90kg).

Shaped for Speed

The pointed head and crescent-shaped tail of the sailfish are found in most high-speed fish. The powerful muscles of the sailfish's long back move the tail from side to side. It does not move up and down as is the case with fast-swimming dolphins.

Other Champions

The swordfish can dart through the waves at 56 miles (91km) an hour. The bluefin tuna is even faster at about 65 miles (104km) an hour. By comparison, Atlantic ocean liners can sail at only 50 miles (80km) an hour at their top speed.

The Snail

One of the slowest animals that moves regularly is the snail. When we talk about slowness, we often refer to snails. The garden snail dashes, or is it plods, along at 0.029 miles (0.048km) per hour. That means it would take it almost 95 years to go around the world! What a pity the snail only lives for 2-3 years.

Snails belong to a group of animals called gastropods. There are over 77,000 of those animals in the world.

The word *gastropod* means "stomach-foot" in Greek and that, in a sense, is what a snail is. It glides on a slimy, muscular foot. At the front of the foot is a snout-like head with a mouth that contains a sort of tongue called a radula. The radula is covered with tiny teeth. The snail moves its radula back and forth over the food. The teeth scrape off small pieces of food to eat.

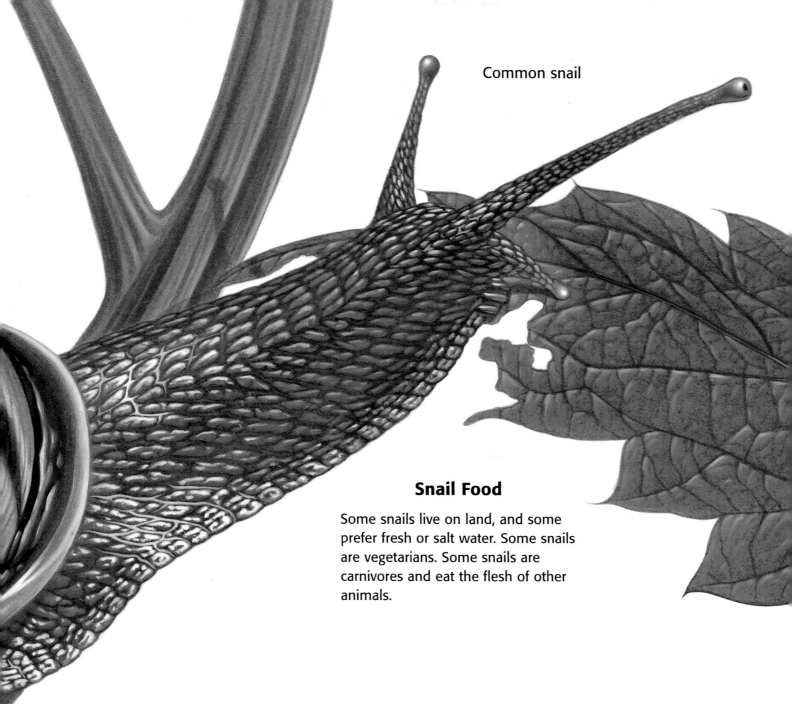

Common snail

Snail Food

Some snails live on land, and some prefer fresh or salt water. Some snails are vegetarians. Some snails are carnivores and eat the flesh of other animals.

Sea Snails

Although most snails are small, a few reach almost giant sizes. The horse conch, found along the Atlantic Coast from North Carolina to Brazil, can have a 24-inch (61-cm) shell. In the Caribbean, the queen conch and king conch make shells almost 12 inches (30cm) long. The animals themselves can weigh up to 5 pounds (2.2kg). The trumpet conch of Australia has a shell that grows to 20 inches (51cm).

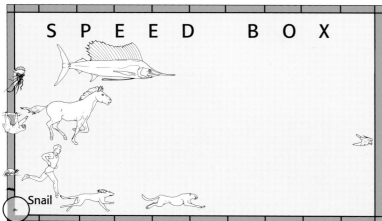

S P E E D   B O X

Snail

0 100 200 300 400 500 600 700 800 900 1000m

The Turtle

For many millions of years turtles have been ambling about. They were around during the age of the dinosaurs and are still around now! They rely on their armor of hard, horny shell for protection. These slow-moving animals are very successful survivors.

Land turtles, or tortoises, are not built for speed. Trying its hardest, a Galapagos tortoise can move at no more than 0.19 mph (0.30kmph) over short distances. Land turtles save their energy as much as possible, and sometimes live very long lives.

The Oldest Tortoises

Although claims have been made that a giant tortoise in the Cairo zoo reached the age of 269, they cannot be confirmed. But tortoises can certainly live up to 120 years. There is at least one reliable case of a tortoise living to be 180 years old.

Flowers and Snails

Tortoises are mainly vegetarians. Galapagos tortoises eat mostly cacti, grasses, leaves of bushes, and fruits. Some other tortoises are fond of flowers, especially those with bright red, orange, or yellow petals. Some tortoises do eat meat and will even hunt for snails, which they gobble down shell and all.

Living Tanks

Tortoises and other turtles are reptiles. This means they have a backbone and watertight, scaly skin. Reptiles also breathe air with lungs. Unlike birds and mammals, reptiles do not have a constant body temperature. Instead, their temperature rises and falls with that of their surroundings. For this reason, they are said to be "cold blooded."

Galapagos tortoise

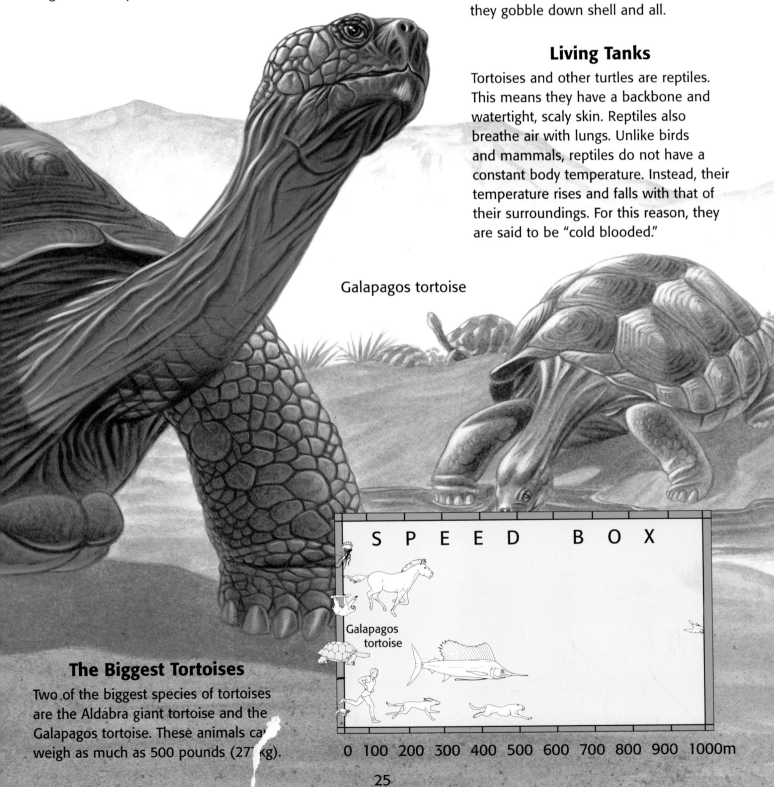

SPEED BOX

Galapagos tortoise

0 100 200 300 400 500 600 700 800 900 1000m

The Biggest Tortoises

Two of the biggest species of tortoises are the Aldabra giant tortoise and the Galapagos tortoise. These animals can weigh as much as 500 pounds (227 kg).

The Peregrine Falcon

Besides fast animals on land and in water, there are those that can shoot through the air. Like the cheetah and the sailfish that need their speed to catch fast-moving prey, some birds can fly very quickly when they are hunting.

One such bird is the Peregrine falcon. Like other birds of prey, such as eagles and hawks, it uses its eyes to find prey below it, as it rides on the **currents** of air. The Peregrine falcon's powerful eyes are two or three times stronger than those of humans. Once a target has been found, the Peregrine folds back its wings and swoops, or dives down on it, striking or seizing it with sharp, curved claws and a hooked beak. During such a dive the falcon can speed up to 150 mph (240kmph).

Are Pigeons Faster?

Peregrine falcons cannot catch up with a racing pigeon flying at its maximum speed of around 89 miles (144km) per hour if the falcon cannot dive down on it from above.

SPEED BOX

Spine-tailed swift

Peregrine falcon

0   100   200   300   400   500   600   700   800   900   1000m

Faster Than a Falcon

The fastest bird of all is the Spine-tailed swift of Asia. This bird has been measured at 105 miles (169km) per hour in level flight, not diving from a great height like a falcon.

Peregrine falcon chasing a spine-tailed swift.

The Killer Whale

Speed, strength, high intelligence—and a mouthful of wickedly pointed teeth. With that combination, it's no wonder that **Inuit** people of North America call the killer whale "Lord of the Sea."

The killer whale is the biggest type of dolphin. Males can grow to a length of 32 feet (10m) and weigh almost 10 tons (9,000kg). Females are slightly smaller, reaching 28 feet (9m) and a weight of 6 tons (5,500kg). Killer whales are afraid of nothing in the sea. They take on sharks, sea lions, leopard seals, dolphins, and big whales. The killer whale can sprint into an attack at speeds over 28 miles per hour (45kmph).

Killer whale

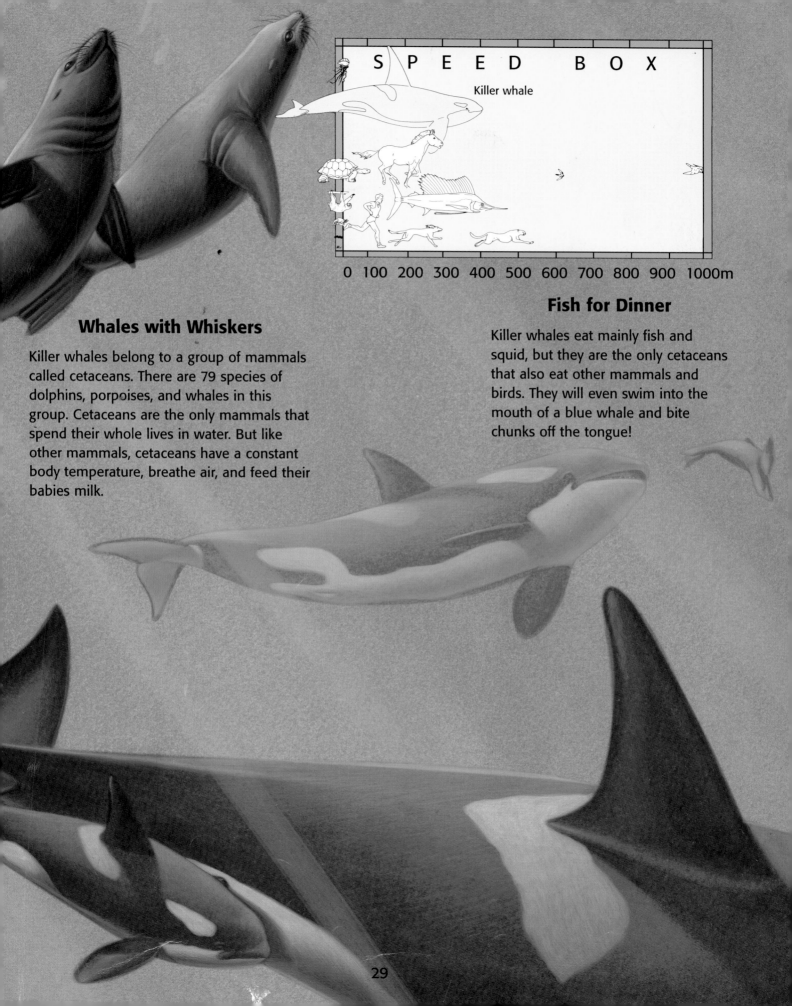

SPEED BOX

Killer whale

0 100 200 300 400 500 600 700 800 900 1000m

Whales with Whiskers

Killer whales belong to a group of mammals called cetaceans. There are 79 species of dolphins, porpoises, and whales in this group. Cetaceans are the only mammals that spend their whole lives in water. But like other mammals, cetaceans have a constant body temperature, breathe air, and feed their babies milk.

Fish for Dinner

Killer whales eat mainly fish and squid, but they are the only cetaceans that also eat other mammals and birds. They will even swim into the mouth of a blue whale and bite chunks off the tongue!

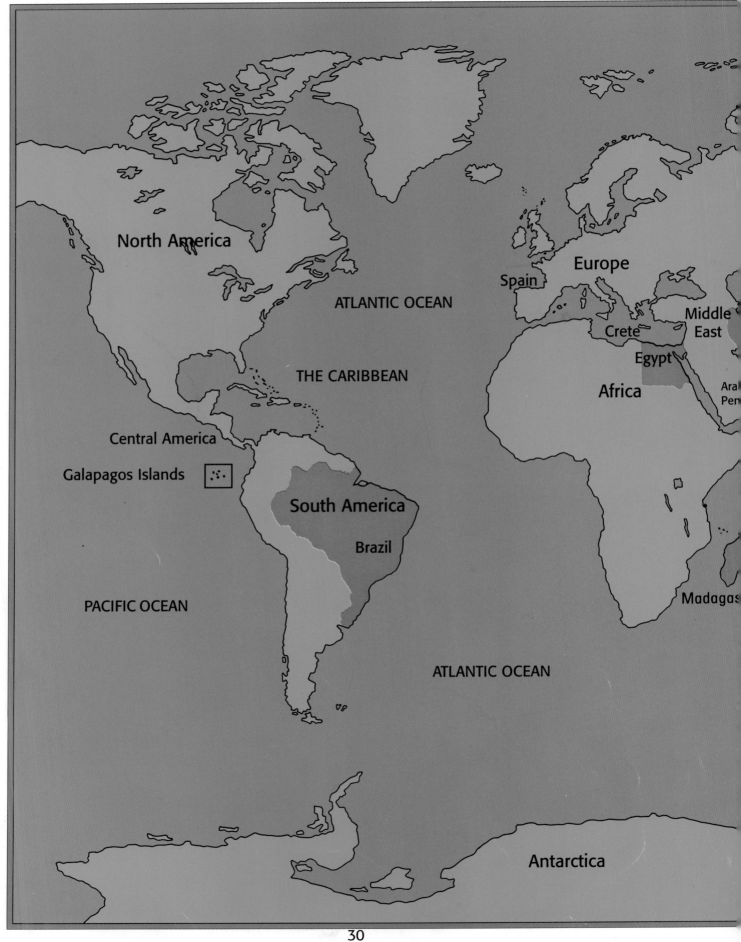

North America

Europe

Spain

ATLANTIC OCEAN

Middle
East

Crete

Egypt

THE CARIBBEAN

Africa

Central America

Galapagos Islands

South America

Brazil

PACIFIC OCEAN

Madagas

ATLANTIC OCEAN

Antarctica

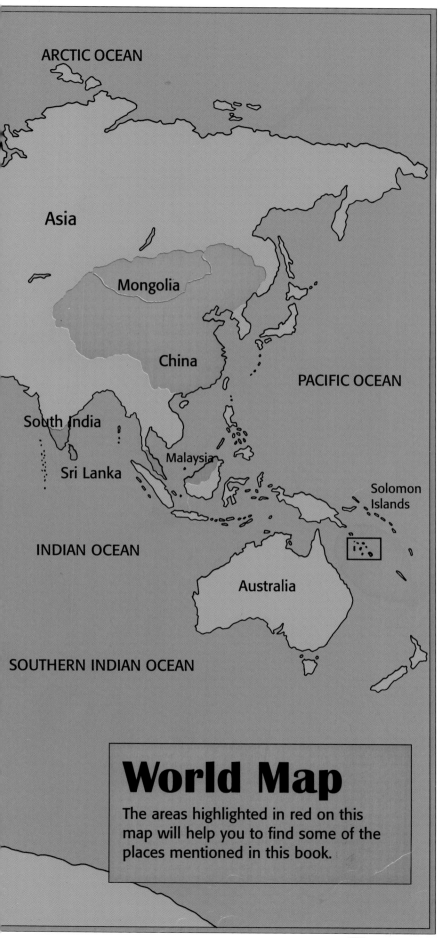

ARCTIC OCEAN

Asia

Mongolia

China

PACIFIC OCEAN

South India

Sri Lanka

Malaysia

Solomon
Islands

INDIAN OCEAN

Australia

SOUTHERN INDIAN OCEAN

World Map

The areas highlighted in red on this map will help you to find some of the places mentioned in this book.

Glossary

Bedouin People of Arabia and Africa who live in and travel through the desert.

chilopods A group of animals without backbones, whose bodies are separated into parts and whose front legs are used as jaws.

currents Air or water that moves in one direction. Animals can be carried along by currents.

domestic Taken from their wild state and tamed over thousands of years by people.

Inuit The Eskimo people who live in North America and Greenland. *Inuit* means "people" or "human beings".

mammal Warm-blooded, hairy animals that feed their young on their own milk. Human beings are mammals.

mummified When a body is preserved after death, it has been mummified. It was normal to mummify the bodies of kings and queens in Ancient Egypt.

orbit The path of an object as it travels around a star or planet.

predators Animals who hunt other animals.

prey Animals who are killed for food by other animals.

species Groups of animals and plants that are similar to each other and have common features. There are between 5 and 50 million species worldwide.

stamina The ability to keep up physical activity for long periods of time.

tentacles Long, thin limbs, for grasping or feeling. These hang down from an animal, such as the jellyfish.

Index

A **bold** number shows the entry is illustrated on that page. A word in **bold** is in the glossary on page 31.